LIZARDS Play Tricks!

BY ANNA COLLINS

ANIMAL TRICKSTERS

Gareth Stevens PUBLISHING

Please visit our website, www.garethstevens.com. For a free color catalog of all our high-quality books, call toll free 1-800-542-2595 or fax 1-877-542-2596.

Cataloging-in-Publication Data

Names: Collins, Anna.
Title: Lizards play tricks! / Anna Collins.
Description: New York : Gareth Stevens Publishing, 2025. | Series: Animal tricksters | Includes glossary and index.
Identifiers: ISBN 9781538293515 (pbk.) | ISBN 9781538293522 (library bound) | ISBN 9781538293539 (ebook)
Subjects: LCSH: Lizards–Juvenile literature. | Lizards–Behavior–Juvenile literature.
Classification: LCC QL666.L2 C65 2025 | DDC 597.95–dc23

Published in 2025 by
Gareth Stevens Publishing
2544 Clinton Street
Buffalo, NY 14224

Designer: Andrea Davison-Bartolotta
Editor: Jennifer Lombardo

Photo credits: Cover, p. 1 Irina Kozorog/Shutterstock.com; series art (background texture) Kriangkrai Namtongbai/Shutterstock.com; series art (lizard skin texture) Vera Larina/Shutterstock.com; p. 5 (bird) Alfmaler/Shutterstock.com; p. 5 (pterosaur, ornithischian dinosaur) GoodStudio/Shutterstock.com; p. 5 (saurischian dinosaur) tinkivinki/Shutterstock.com; p. 5 (snake, lizard) Maquiladora/Shutterstock.com; pp. 7, 8, 9 Sergey Uryadnikov/Shutterstock.com; p. 10 reptiles4all/Shutterstock.com; p. 11, 15 (top) Milan Zygmunt/Shutterstock.com; p. 12 IrinaK/Shutterstock.com; p. 13 dwi putra stock/Shutterstock.com; pp. 15 (bottom), 17 (top) Kurit afshen/Shutterstock.com; p. 16 Cathy Keifer/Shutterstock.com; p. 17 (bottom) Manick.Jr/Shutterstock.com; p. 18 Eric Isselee/Shutterstock.com; p. 19 Ken Griffiths/Shutterstock.com; p. 21 Matt Jeppson/Shutterstock.com.

Printed in the United States of America

CPSIA compliance information: Batch #CSGS25: For further information contact Gareth Stevens at 1-800-542-2595.

CONTENTS

Words in the glossary appear in **bold** type the first time they are used in the text.

A CLOSER LOOK AT LIZARDS

Lizards are **reptiles** that are closely related to snakes. There are more than 5,500 different species, or kinds, of lizards! Lizards live all over the world. However, most are happiest in warm places.

Some lizards look so different from each other that you might not even realize they're all related! The biggest ones can weigh more than 150 pounds (68 kg), and the smallest can weigh less than 0.02 ounce (0.5 g). Many lizards have cool survival tricks.

FUN FACT

Scientists say snakes are lizards without legs!

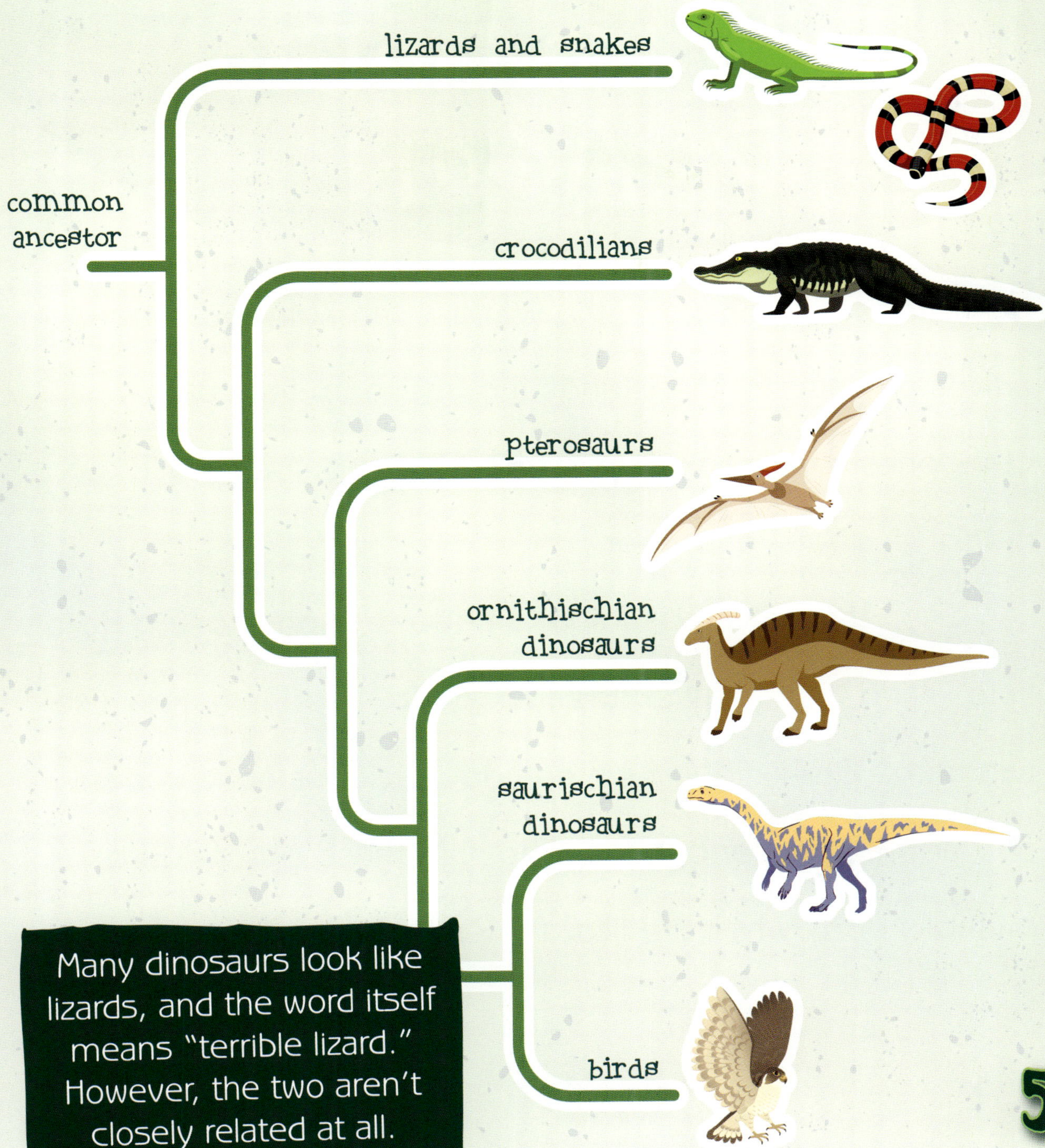

Many dinosaurs look like lizards, and the word itself means "terrible lizard." However, the two aren't closely related at all.

THE KOMODO DRAGON

The biggest lizard in the world is the Komodo dragon. It lives on some of the islands that make up the country of Indonesia. Adult Komodo dragons eat animals such as monkeys, goats, and deer.

Even though they're big, Komodo dragons have a hard time killing their **prey** on the first try. To help them, they use two tricks. One is their bite. Komodo dragons have **venom** and harmful bacteria in their mouths. These get inside an animal's body when the lizard bites it.

The Komodo dragon is one of the only lizards in the world whose bite can be deadly to humans.

FUN FACT

Most Komodo dragons weigh around 150 pounds (68 kg). However, the largest one ever found weighed 366 pounds (166 kg) and was 10.3 feet (3.1 m) long!

ON THE HUNT

The Komodo dragon's venom and bacteria don't kill its prey right away. In fact, it can take up to four days for an animal to die after it's bitten. This gives it a lot of time to get away. A Komodo dragon's second trick is its powerful sense of smell.

As the prey runs away, the Komodo dragon walks slowly after it. It swings its head from side to side, tasting the air as it goes. It follows the smell until it finds its prey's dead body.

FUN FACT

A Komodo dragon can smell rotting meat up to 2.5 miles (4 km) away.

Komodo dragons smell by tasting the air with their tongue.

THE GILA MONSTER

There aren't many venomous lizards in the world. Aside from the Komodo dragon, the best-known is the Gila monster. This type of lizard has black and pink markings on its body. It is the largest lizard in the United States.

Most venomous animals have hollow fangs, or pointed teeth, that they push venom through when they bite. However, instead of being hollow, the Gila monster's teeth have grooves. Venom flows through these **grooves** into their prey.

Gila monsters live in deserts, mainly in the southwest part of the United States.

FUN FACT

A Gila monster's venom is about as powerful as the venom of a western diamondback rattlesnake.

THE BLUE-TAILED SKINK

The Komodo dragon and Gila monster have tricks that help them catch prey. Other lizards have tricks that help them avoid being caught! The blue-tailed skink is named for its bright blue tail. The color attracts the skink's predators, such as birds, foxes, skunks, snakes, and cats.

When a skink is scared, its tail falls off and **twitches** on the ground. The tail's color and movement catch the predator's attention. This gives the skink time to run away. Its tail grows back later.

Blue-tailed skinks are mainly found in forests.

THE CHAMELEON

The chameleon is another colorful lizard. In fact, it can change its color! One reason why chameleons do this is to make themselves warmer or cooler. Darker colors **absorb** more heat, while lighter colors **reflect** it.

Chameleons also change color based on their feelings. When they want to fight or **mate** with another chameleon, they tend to turn bright red or yellow. When they want to show that they don't want to fight, they tend to make their skin look darker.

Chameleons change color to show off, not to hide. When they're calm, they naturally match the sky, trees, or ground.

AMAZING EYES

The chameleon's tricks aren't limited to color. It also has eyes unlike any other animal's. Instead of an upper and lower eyelid, a chameleon has one big eyelid shaped like a cone. In the center of this cone is a small opening for the chameleon to look out.

The eyes can move in almost any direction, and they can move independently of each other. When a chameleon spots something, its eyes can zoom in like a camera lens to see whether it's food, a predator, or something else.

When a chameleon spots food, it shoots out its long, sticky tongue to grab it.

THE FRILLED-NECK LIZARD

The frilled-neck lizard is well known in its homeland of Australia. It gets its name from the large piece of skin, called a frill, around its neck. When it feels scared, the lizard can raise its frill to make itself look bigger than it is. Predators don't want to get in a fight with a big animal!

If a predator isn't scared off by the frill, the lizard has another trick: It can raise itself up on its back legs to help it run away faster.

This frilled-neck lizard is warning away a possible predator.

THE TEXAS HORNED LIZARD

There are many different species of horned lizard. One of them, the Texas horned lizard, has many different tricks it can use to stay safe from predators. One of its tricks is that it can inflate, or puff up, to make itself look bigger. Sometimes this is enough to scare away predators.

If inflating itself doesn't work, the Texas horned lizard has an even more amazing trick up its sleeve: It can shoot blood from its eyelids! The blood tastes bad to most of the lizard's predators. They forget about the lizard while they try to get the taste out of their mouth.

The Texas horned lizard's sharp **spines** make it hard for most predators to eat.

FUN FACT

Not all horned lizards can shoot blood from their eyes. Out of 22 species, only eight are known to do this trick.

GLOSSARY

absorb: To take in.

groove: A narrow space through which liquid can flow.

mate: When two animals come together to make babies.

prey: An animal that is hunted by other animals for food.

reflect: To give back light or heat.

reptile: An animal covered with scales or plates that breathes air, has a backbone, and lays eggs, such as a turtle, snake, lizard, or crocodile.

spines: A stiff, sharp part of a plant or animal.

twitch: A quick, sudden movement.

venom: Something an animal makes in its body that can harm other animals, including humans.

FOR MORE INFORMATION

Books

Klepeis, Alicia. *Chameleons.* Minneapolis, MN: Jump!, Inc., 2021.

Munro, Roxie. *Lizards at Large: 21 Remarkable Reptiles at Their Actual Size.* New York, NY: Holiday House, 2023.

Nicholas, Christopher. *Lizards.* London, UK: Sequoia Kids Media, 2023.

Websites

BrainPOP: Reptiles
www.brainpop.com/science/diversityoflife/reptiles
Learn about lizards and other reptiles through games, videos, and quizzes.

Kiddle: Lizard Facts for Kids
kids.kiddle.co/Lizard
Read more facts about lizards and take a look at some cool pictures.

World Wildlife Fund: Ten Fierce Facts About Komodo Dragons
www.worldwildlife.org/blogs/good-nature-travel/posts/ten-fierce-facts-about-komodo-dragons
This website gives more information about the world's largest lizard.

Publisher's note to educators and parents: Our editors have carefully reviewed these websites to ensure that they are suitable for students. Many websites change frequently, however, and we cannot guarantee that a site's future contents will continue to meet our high standards of quality and educational value. Be advised that students should be closely supervised whenever they access the internet.

INDEX